Introduction to Geometry Made Simple

MW00963669

Written by Theresa Kane McKell

Illustrated by Don O'Connor

Cover Design by Signature Design Group

FS122011 Introduction to Geometry Made Simple
All rights reserved—Printed in the U.S.A.
Copyright © 1999 Frank Schaffer Publications, Inc.
23740 Hawthorne Blvd.
Torrance, CA 90505

Introduction

Geometry plays an important part in students' understanding of the world around them. Geometry organizes all shapes and their properties that students interact with. It is essential that students learn to develop, understand, and learn to apply geometric skills.

What is a rectangle?

How will geometry make me better understand manufacturing, painting, and installing?

How are 3-D and 2-D objects related?

How is surface area and volume computed?

Answering these and other questions are students' attempts to understand geometric concepts.

Introduction to Geometry Made Simple has been written to aid students in the development of a basic understanding of geometric concepts and to help them learn and practice the skills necessary for this type of understanding. It contains activities that have been designed to provide fun and exciting ways to learn geometry and to give students a variety of everyday applications.

The objective of *Introduction to Geometry Made Simple* is to give all students the opportunity to experience success in geometry. To help ensure this success, a wonderful variety of activities have been included and encompass a variety of different constructions, picture sketches, and puzzles, some relating to real-life records, dates, and facts, others featuring riddlelike situations.

....

This book is divided into six sections. At the beginning of each section are teacher resource pages. These pages contain many related activities and problems that can be used to guide the students through each section. One of the exciting aspects of these teacher resource pages is the group activities. These activities provide a fun way to help students work together to apply the concepts and skills presented in the section in order to create steps to solve a particular situation. This kind of group work allows students to gain the experience of working with other students and gives them the opportunity to see the real-life applications of math in the world around them.

Following each set of teacher resource pages are interesting and exciting student activity pages. Students can work these pages to practice their skills and gain a conceptual understanding of the topics in each particular section. Some of the activities involve students in constructing geometric figures, working puzzles, solving riddles, and decoding messages. And while the primary focus of each activity is the featured geometry skill, students will also enjoy the many jokes, record breakers, and riddles included.

....

The concepts covered in *Introduction to Geometry Made Simple* are basic to most middle school math curriculums. Students will develop a conceptual understanding of the geometric topics presented and will practice skills relating to the following concepts: points, lines, line segments, rays, planes, and angles; characteristics of polygons; triangles and quadrilaterals; perimeter and circumference; area of polygons; and surface area and volume of three-dimensional figures.

This is the perfect opportunity to make learning geometry fun, relevant, and interesting for any student. *Introduction to Geometry Made Simple* is an easy way to develop students' interest in and understanding of valuable geometric concepts. You will be excited to observe as your students discover how stimulating learning geometric concepts can be!

FS122011 Introduction to Geometry Made Simple ▪ © Frank Schaffer Publications, Inc.

Points, Lines, Line Segments, Rays, Planes, and Angles

Every math student will greatly benefit from the activities involving the basic ideas of geometry in this section. Allow students ample opportunity to work with manipulatives and time to complete several examples with your guidance. Be sure students gain a conceptual understanding of the concepts to the right before proceeding through the independent student activity pages (pages 3–18).

Present everyday situations to students in which they may use their new skills. For example, students can use their knowledge of the basic ideas of geometry when manufacturing special products, connecting cities to cities, and when identifying the exact location of something. Help students observe the world in which they live and identify their own connections involving working with the basic ideas of geometry.

CONCEPTS

The ideas and activities presented in this section will help students explore the following concepts:

- naming and drawing points, lines, line segments, and rays
- naming and drawing parallel and perpendicular lines
- naming and drawing intersecting and skew lines
- working with parallel lines and transversals
- naming and drawing planes and angles
- constructing segments and angles
- constructing perpendicular, parallel, and intersecting lines
- naming acute, right, and obtuse angles
- finding complementary and supplementary angles
- finding vertical and adjacent angles
- finding angle measures
- bisecting segments and angles

GETTING STARTED

- Ask students to lay a straightedge where it touches a Point A in several different ways. Using this as a demonstration, ask them to define exactly how many lines could pass through that same Point A. (infinitely many lines) Give students two points, A and B. Ask them how many lines can pass through <u>both</u> of these points together. (only one) Ask them what this says about the number of lines determined by exactly two points. (only one line can pass through two points)

- Have students experiment with intersecting lines using two pencils. Ask them to show that the following statement about two intersecting lines is true no matter how they position the pencils: *They will always only touch at one point.*

- To emphasize the specific distinctions between lines, segments, and rays, ask students to draw line AB, segment AB, and ray AB.

- Ask students to name the angles to the right in as many ways as possible. (∠ABC, ∠CBA, ∠1, ∠B; ∠DEF, ∠FED, ∠2, ∠E)

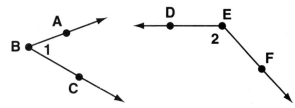

GEOMETRY IN EVERYDAY LIFE

Draw a picture of a picnic table, scissors, school desk, and telephone pole on the board. Ask students to name the parts in the pictures that represent points; lines; line segments; intersecting, parallel, and perpendicular lines; rays; planes; or angles.

ANGLE INVESTIGATION

Ask students to draw several pairs of intersecting lines. Have them label all angles formed and name all pairs of vertical angles. Ask them to use this to determine exactly how many pairs of vertical angles are formed by a pair of intersecting lines. You may extend the same activity for the number of pairs of supplementary angles formed by a pair of intersecting lines.

ANGLE REPRESENTATION

The hands on a clock represent an angle. Ask students to name the angle measure formed at each hour below.

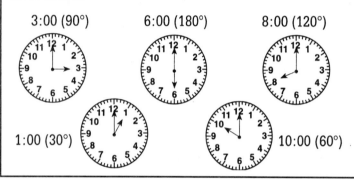

3:00 (90°)   6:00 (180°)   8:00 (120°)

1:00 (30°)   10:00 (60°)

CONGRUENCE QUIZ

Draw the figures below on the board and have students name the pairs of figures that are congruent. (a ≅ f, b ≅ d, c ≅ e)

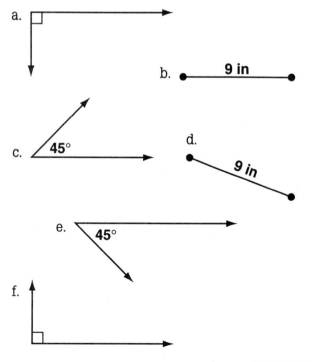

a.

b. •——9 in——•

c. 45°

d.

9 in

e. 45°

f.

QUICK QUIZ

Ask students to state whether each statement below is true or false. If false, have them explain why.

- Only one line can be drawn through two points. (T)
- An angle is made up of two rays joined at the vertex. (T)
- Given exactly one point, exactly one line can be drawn through that point. (F—infinitely many can)
- Two lines can intersect at infinitely many points. (F—exactly one point)
- If given three points on a line, you will have three different segments. (T)
- If the sum of two angles is 180°, then the angles are complementary. (F—supplementary angles)
- Adjacent angles will always have the same measure. (F—sometimes)

Group Activity

Divide students into groups of three. Give each group four different objects from the classroom. Have each group estimate the lengths of all sides of each object. Then have them accurately measure the sides to get the precise lengths. Ask them to compare the actual measures of the sides and the estimated measures. How far off were they? Ask them to explain the methods they used to estimate and, if they were way off from the actual measurements, explain why. Ask each group to switch objects with another group and repeat the same activity. When finished with this set of objects, have each group compare its results with the group it had switched with. Have them explain the similarities and differences in their measurements and estimations. What are some of the reasons there could be discrepancies in these measurements?

Basic Training

Using symbols, name the drawings below in as many ways as possible.

1.

• A

2.

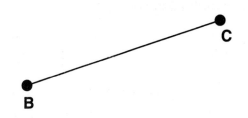

3.

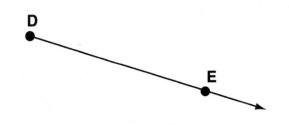

4.

5.

6.

7.

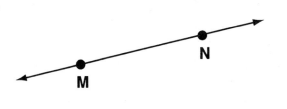

8.

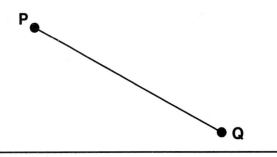

9.

10.

Drawing Doodles

Draw and label each of the problems below.

1. line segment PQ	**2.** point Z
3. line XY	**4.** ray AB
5. point M on line RS	**6.** points C and D on ray TU
7. line segment GH with point O between points G and H	**8.** point L on line segment JK
9. point E between points R and T on line VW	**10.** ray XY with point B between points X and Y

The Dancing Computer

Where would a computer go if it wanted to dance?

To find out, identify whether each of the pairs of lines below is parallel or perpendicular. Circle the letter that represents the correct answer. Write each letter above its corresponding problem number at the bottom of the page.

1. (T) parallel   (G) perpendicular

2. (I) parallel   (O) perpendicular

3. (S) parallel   (A) perpendicular

4. (D) parallel   (Y) perpendicular

5. (I) parallel   (U) perpendicular

6. (H) parallel   (S) perpendicular

7. (G) parallel   (K) perpendicular

8. (O) parallel   (R) perpendicular

___ ___   ___   ___ ___ ___ ___   –   ___
 1   2   3   4   5   6   7   8

Double Draw Dare

Draw and label each of the problems below.

1. Line AB is parallel to line RS.	**2.** Line segment TU is perpendicular to line segment CD.
3. Ray GH is perpendicular to line MN at point W.	**4.** Line XY is perpendicular to both parallel lines UV and ST.
5. Line segment OP is parallel to ray MN.	**6.** Line JK is perpendicular to line segment QR at point I.
7. Ray OH is parallel to line TM and line segment UV.	**8.** Line MN is perpendicular to both parallel lines EF and CD at points H and W, respectively.

FS122011 Introduction to Geometry Made Simple ▪ © Frank Schaffer Publications, Inc.

State of Combination

About 165 years ago, what was the first state in the United States to allow boys and girls to be taught in the same classes?

To find out, identify whether each pair of lines below is intersecting or skew. Circle the letter that represents the correct answer. Read the circled letters downward to identify the answer.

1. 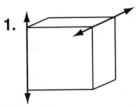　　　(M) intersecting　　　(I) skew

2. 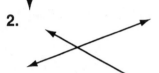　　　(N) intersecting　　　(O) skew

3. 　　　(D) intersecting　　　(N) skew

4. 　　　(I) intersecting　　　(T) skew

5. 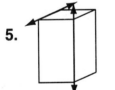　　　(L) intersecting　　　(A) skew

6. 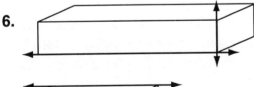　　　(N) intersecting　　　(M) skew

7. 　　　(O) intersecting　　　(A) skew

Answer: _____

Intersecting vs. Skew

Draw and label each of the problems below.

1. Line ST and line RY intersect.

2. Lines AB and CD are skew lines.

3. Ray FG intersects line segment HI.

4. Line segments BC and UV are skew segments.

5. Line RT intersects line XY at point M.

6. Ray JK intersects lines UI and MN at points Y and L, respectively.

7. Line WX and line segment CG are skew.

8. Line segment TS intersects lines RH, JK, and NM at points U, O, and E, respectively.

FS122011 Introduction to Geometry Made Simple ▪ © Frank Schaffer Publications, Inc.

Riveting River

What is the name of the longest river in the world, extending 4,145 miles long?

To find out, find the measure of angles 1–7 using the given information. To spell out the answer to the riddle, write the letter of the corresponding problem above the given answer.

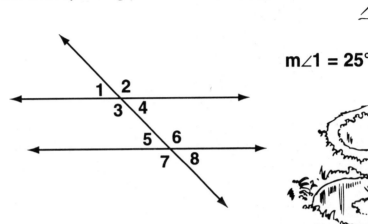

m∠1 = 25°

E. m∠2 = _____

I. m∠4 = _____

E. m∠3 + m∠4 = _____

N. m∠5 + m∠4 + m∠3 = _____

T. m∠6 + m∠7 = _____

L. m∠5 + m∠8 = _____

H. m∠6 – m∠8 = _____

___	___	___		___	___	___	___
310°	130°	180°		205°	25°	50°	155°

Symbolic Names

Using symbols, name each of the drawings below two different ways. Write your answers in the blanks next to each problem number.

1. _____
_____

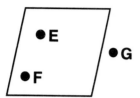

2. _____
_____

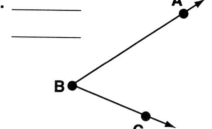

3. _____
_____

4. _____
_____

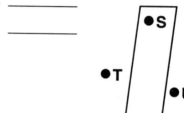

5. _____
_____

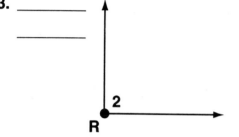

6. _____
_____

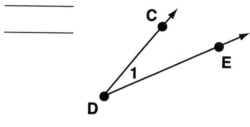

7. _____
_____

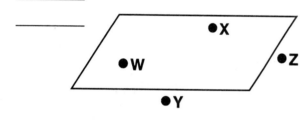

8. _____
_____

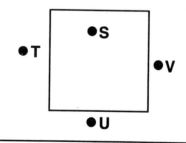

9. _____
_____

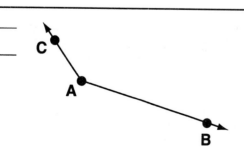

10. _____
_____

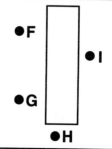

FS122011 Introduction to Geometry Made Simple ▪ © Frank Schaffer Publications, Inc.

Dare to Design

Draw and label each of the problems below.

1. angle HIJ	**2.** angle ABC with points X and Y on the angle
3. plane MNO	**4.** line GH in plane Z
5. points T and Y in plane R	**6.** angle XYZ in plane C
7. plane W	**8.** angle OHS in plane FGH

Construction Site

Use a compass and a straightedge to construct a figure congruent to each line segment and angle drawn below. Draw your constructions to the right of each existing figure.

CONGRUENT CONSTRUCTION

1.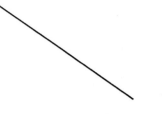

2.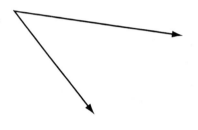

3.

4.

5.

6.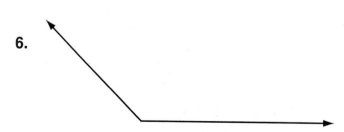

FS122011 Introduction to Geometry Made Simple ▪ © Frank Schaffer Publications, Inc.

Mindful Construction

Use a compass and a straightedge to construct the lines below.

1. line XY that is parallel to line RS

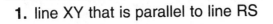

R   S

2. the perpendicular bisector, line TH, of line MN—Label their intersection at point A.

M   N

3. the perpendicular bisector, line GH, of parallel lines BN and AJ—Label the points of intersection at points Y and Z, respectively.

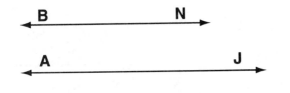

B   N

A   J

4. line AB that is parallel to line XY with a perpendicular bisector (line ER)

X   Y

Young Intelligence

Michael Kearney is the youngest person in the world to ever have attended a university while working on his Associate of Science degree at Santa Rosa Junior College in Santa Rosa, California. How old was he?

To find out, measure each angle below using a protractor. Classify each angle as acute, obtuse, or right. Write the angle measure on the blank inside the angle and its classification on the blank next to its problem number. Your total number of obtuse angles will give you the answer to the amazing question.

1. _____

2. _____
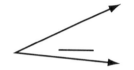

3. _____

4. _____
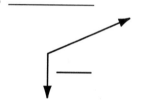

5. _____
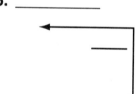

6. _____
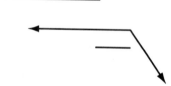

7. _____
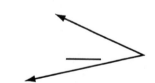

8. _____

9. _____
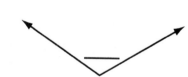

10. _____
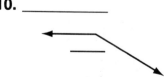

11. _____
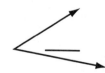

12. _____
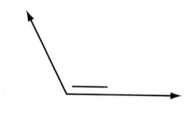

Answer: _____

FS122011 Introduction to Geometry Made Simple ▪ © Frank Schaffer Publications, Inc.

Name_____

Lavish Lawyers

What do lawyers usually wear when they go to work?

To find out, find the measure of each of the missing angles. Find your solutions in the list of answers provided. Write the letters by the corresponding problem numbers to spell out the answer.

1. _____

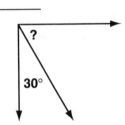

2. _____

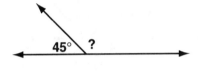

3. _____

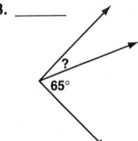

4. _____

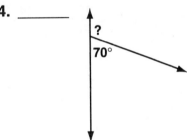

5. _____

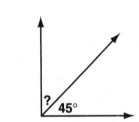

6. _____

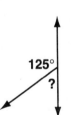

7. _____

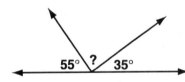

8. _____

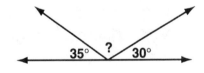

A	J	T	K	S	Y
135°	65°	90°	30°	115°	15°
E	**S**	**W**	**U**	**L**	**I**
75°	110°	25°	45°	60°	55°

Answer: _____

Angle-Mazing

Use the drawing below to answer the questions relating to vertical and adjacent angles.

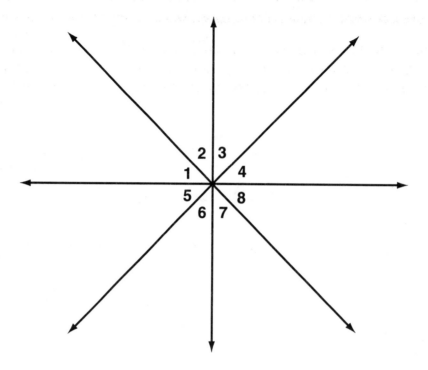

1. Name all pairs of adjacent angles.

_____

2. Name all pairs of vertical angles.

_____

3. If m∠1 = 40°, what is the measure of ∠8? _____

4. The sum of angles 6 and 7 is equal to the sum of what two angles?_____

5. If m∠4 + m∠8 = 80°, what does the sum of m∠1 and m∠5 equal? _____

6. Name the vertical and adjacent angles of ∠3.

_____

7. If m∠3 = 50°, what is the measure of ∠6? _____

8. The sum of angles 1, 3, and 7 is going to equal the sum of what three angles?

_____

   FS122011 Introduction to Geometry Made Simple ▪ © Frank Schaffer Publications, Inc.

Mammoth Mammal

What holds the record for the largest mammal, even the largest animal, ever recorded?

To find out, use your knowledge of vertical, adjacent, complementary, and supplementary angles and the figure below to find the missing measures. Shade in the boxes that contain your answers. Read across the remaining unshaded boxes to spell out the answer to the record-holding question.

1. $m\angle 6$ = _____

2. $m\angle 4$ = _____

3. $m\angle 2$ = _____

4. $m\angle 6 + m\angle 5 + m\angle 4$ = _____

5. $m\angle 1 + m\angle 3$ = _____

6. $m\angle 2 - m\angle 5$ = _____

7. $m\angle 5 - m\angle 1$ = _____

8. $m\angle 1 + m\angle 2 + m\angle 3 + m\angle 4$ = _____

9. $m\angle 1 + m\angle 2 + m\angle 3 + m\angle 4 + m\angle 5 + m\angle 6$ = _____

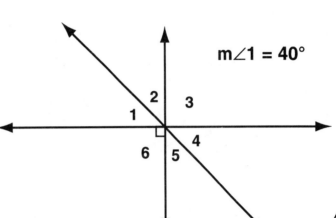

$m\angle 1 = 40°$

H	B	I	L	P	K
50°	55°	0°	15°	360°	90°
U	**Y**	**T**	**E**	**W**	**H**
65°	130°	10°	25°	95°	45°
N	**A**	**M**	**L**	**R**	**E**
40°	145°	220°	150°	180°	60°

Answer: _____

Bisecting Babble

Use a compass and a straightedge to bisect the line segments and angles below.

1.
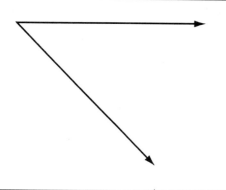

2.
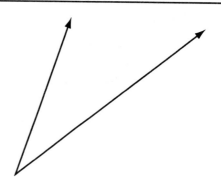

3.

4.

5.
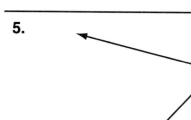

6.
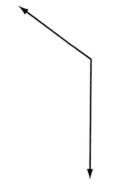

7.
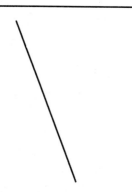

8.

FS122011 Introduction to Geometry Made Simple ▪ © Frank Schaffer Publications, Inc.

Special Characteristics of Polygons

Every math student will greatly benefit from the activities involving special characteristics of polygons in this section. Allow students ample opportunity to work with manipulatives and time to complete several examples with your guidance. Be sure students gain a conceptual understanding of the concepts to the right before proceeding through the independent student activity pages (pages 21–31).

Present everyday situations to students in which they may use their new skills. For example, students can use their knowledge of special characteristics of polygons when installing a material in a room in their house, when painting a wall or some other surface, and when calculating proportionality in construction. Help students observe the world in which they live and identify their own connections involving working with special characteristics of polygons.

CONCEPTS

The ideas and activities presented in this section will help students explore the following concepts:
- sum of angle measures of polygons
- number of sides, angles, and diagonals of polygons
- regular polygons
- convex and concave polygons
- lines of symmetry
- turns of symmetry
- congruent polygons
- interior angles of polygons

GETTING STARTED

Draw a square, rectangle, parallelogram, and triangle on the board. Give the related dimensions for each shape as shown below.

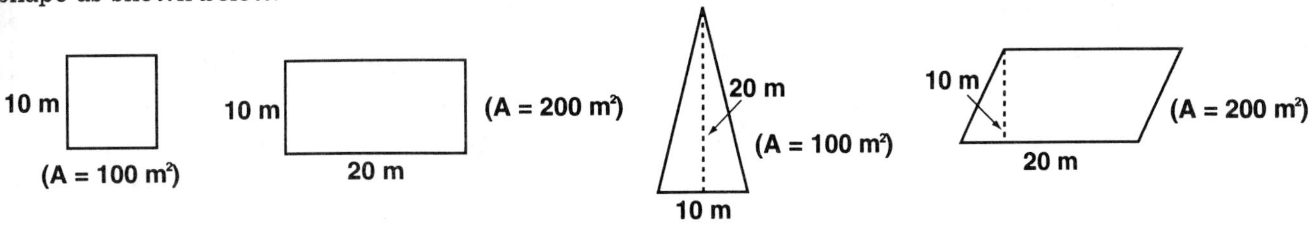

Ask students to compare the areas of these figures. How do they relate to one another? Is there any conjecture that can be made for the relationship of these areas? (area of triangle is half the area of the parallelogram with the same dimensions) What has to be true about the dimensions of the shapes for such a relationship to occur? (They have to be the same.) Have students explain in detail their reasoning behind their conclusions.

PARTNER FUN

Divide students into pairs. Give each of them a geoboard. Have each pair list six different polygons. Ask students to use rubber bands and their geoboards to create the six polygons. Have them draw the shapes on paper and name them accordingly. Have them model each polygon on the geoboards in as many ways as possible.

POLYGON INVESTIGATION

Ask students to draw regular polygons with 3, 4, and 5 sides. Have them draw all diagonals. Ask them to set up a table with the name of the polygon, the number of its sides, the number of its angles, and the number of its diagonals. Have students use this information to complete the table with a regular hexagon, regular heptagon, regular octagon, and regular decagon. What is the pattern of finding the number of diagonals in a regular polygon? [$n(n-3)/2$] Be sure students explain their answers in detail.

SHAPE UP OR SHIP OUT

Draw each shape below on the board. Ask students to tell you why each one is not a polygon. Have them explain their reasoning.

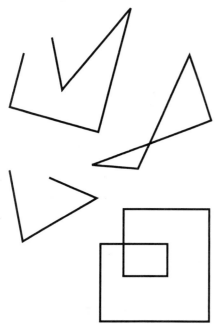

QUICK QUIZ

Ask students to state whether each statement below is true or false. If false, have them explain why.

- A diagonal of a polygon is a line segment that connects two vertices that are adjacent. (F—connects 2 non-adjacent vertices)
- Regular polygons have all sides with the same length and all angles with the same measure. (T)
- Convex polygons have at least one interior angle that is greater than 180°. (F—concave)
- Two polygons are congruent if they have the same size but not necessarily the same shape. (F—same size and same shape)
- A square has two lines of symmetry and four turns of symmetry. (F—4 lines of symmetry and 90° turn of symmetry)

Group Activity

Divide students into groups of three or four. Ask them to imagine the following problem and solve it accordingly:

New ceramic tiling is needed for several classrooms in the school. Each room that needs new tiling is rectangular. Three rooms on each floor need to be tiled, except for the top floor, where only 2 rooms need tiling. The total number of rooms that need to be tiled is a prime number. (Answer: 4 floors, 11 rooms)

Ask students to work in their groups to make some conjectures based on the information given. Have them write these down and be able to explain each to the class. Ask them to answer the questions below based on the given information.

- If each classroom is rectangular, how many sides does each of the rooms that need tiling have?
- If a coat room in one of the rooms is 64 square feet, what are some of the possible shapes of the coat room?
- What shapes could the tiles be if we know the rooms are rectangular?
- Assuming that your school has more than two floors, how many classrooms does the school have?

Have students use this information to conclude the number of rooms to be tiled. They should also explain their reasoning.

FS122011 Introduction to Geometry Made Simple ■ © Frank Schaffer Publications, Inc.

Carpenter Craze

What did the crazy carpenter do every night before he went to bed?

To find out, find the sum of the measures of each polygon below. Write the problem number in front of the corresponding answer listed in the table. To spell out the answer at the bottom of the page, refer to the table and write the code letter that corresponds to the problem number.

1.

Code Letter	Problem #	Answer
A		180°
B		3240°
D		360°
E		1080°
H		2520°
I		540°
M		1800°
S		720°

2.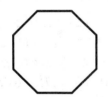

3.

4.

5.

6. a 12-sided polygon

7. a 16-sided polygon

8. a 20-sided polygon

___ ___   ___ ___ ___ ___   ___ ___ ___   ___ ___ ___ !
 7   3   6   5   1   3   7   2   4   8   3   1

Clouds of Shapes

Below is a sky full of shapely clouds. In each cloud, you will find a polygon. Fill in the blanks under each cloud with the name of the polygon (according to its number of sides) and its number of sides, angles, and diagonals.

1.

Name: _____

sides = _____   angles = _____

diagonals = _____

2.

Name: _____

sides = _____   angles = _____

diagonals = _____

3.

Name: _____

sides = _____   angles = _____

diagonals = _____

4.

Name: _____

sides = _____   angles = _____

diagonals = _____

5.

Name: _____

sides = _____   angles = _____

diagonals = _____

6.

Name: _____

sides = _____   angles = _____

diagonals = _____

    FS122011 Introduction to Geometry Made Simple • © Frank Schaffer Publications, Inc.

Massive Munchies

The largest one of these ever made weighed 3,739 pounds. It had a diameter of 16 feet and a height of 16 inches. What was this awesome treat we all love to eat?

To find out, identify the name of each polygon according to the number of its sides. Match each shape in Column A to its name in Column B. Read down the column of written letters to discover the two-word answer to this world-record question.

COLUMN A

____ 1.

____ 2.

____ 3.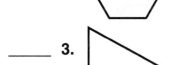

____ 4.

____ 5.

____ 6.

____ 7.

____ 8.

____ 9.

COLUMN B

U. dodecagon

H. decagon

T. nonagon

O. triangle

A. quadrilateral

G. pentagon

U. heptagon

N. octagon

D. hexagon

Answer: _____

Cartoon Character

What is the name of the longest running prime-time animated show in television history?

To find out, classify each shape below. Circle the letters that represent the correct characteristics. Unscramble the circled letters to spell out the name of this record-making TV series.

1.    (J) regular   (M) not a polygon

2. 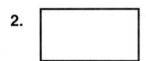   (E) regular   (O) not regular

3. 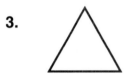   (S) regular   (I) not regular

4.    (N) regular   (I) not regular

5.    (N) regular   (F) not regular

6.    (E) regular   (S) not a polygon

7.    (I) regular   (P) not regular

8. 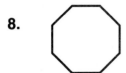   (S) regular   (D) not regular

Answer: "The _____"

Barbecue Bash

Where do monkeys barbecue their meat?

To find out, classify each polygon below with the appropriate characteristic. Circle the letters that represent the characteristics. Place the circled letters in the blanks at the bottom of the page above the corresponding problem numbers.

1. 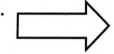   (O) concave   (W) convex

2.   (N) concave   (I)   convex

3.   (T) concave   (G) convex

4.   (H) concave   (R) not a polygon

5.   (T) concave   (I)   convex

6.   (L) not a polygon   (O) concave

7.   (N) convex   (L) not a polygon

8.   (A) concave   (G) convex

9.   (S) not a polygon   (E) concave

___ ___   ___ ___ ___ ___ ___ ___ ___
 1   2   3   4   5   6   7   8   9

Polygon Pictures

Create each drawing below. Name each polygon according to the number of its sides. Write each answer in the blank next to its problem number.

_____ **1.** eight-sided regular polygon

_____ **2.** five-sided convex polygon

_____ **3.** ten-sided concave polygon

_____ **4.** six-sided regular polygon

_____ **5.** not regular seven-sided polygon

_____ **6.** four-sided regular convex polygon

FS122011 Introduction to Geometry Made Simple ▪ © Frank Schaffer Publications, Inc.

Golf Great

In 1997, Tiger Woods became the youngest golfer ever to win the Masters tournament. How old was he?

To find out, draw all lines of symmetry in each figure below. Write the number of lines of symmetry in the blank next to each figure. Add up the total number of lines of symmetry to find the answer to the world record set by Tiger.

_____ 1.

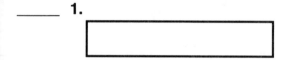

_____ 2.

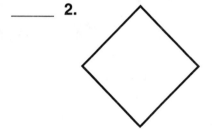

_____ 3.

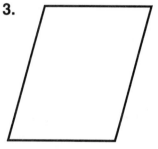

_____ 4.

_____ 5.

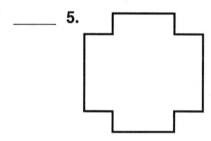

_____ 6.

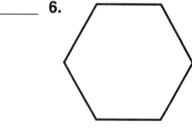

_____ 7.

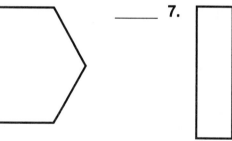

_____ 8.

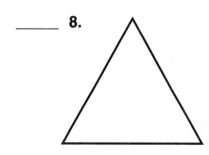

_____ 9.

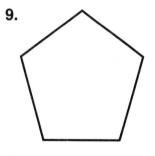

_____ 10.

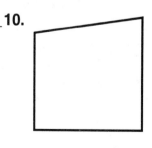

Answer: _____

Grand Slam Hitters

Eight players hold the record for the most grand slams hit in a single major league baseball game. How many grand slams is this?

To find out, identify the angle measure of each turn of symmetry in the problems below. Write your answer in the blank next to each problem number. If the shape has no turn of symmetry, write "no" in the blank. Count the total number of "no" answers—this will give you the number of the most grand slams ever hit in a single game.

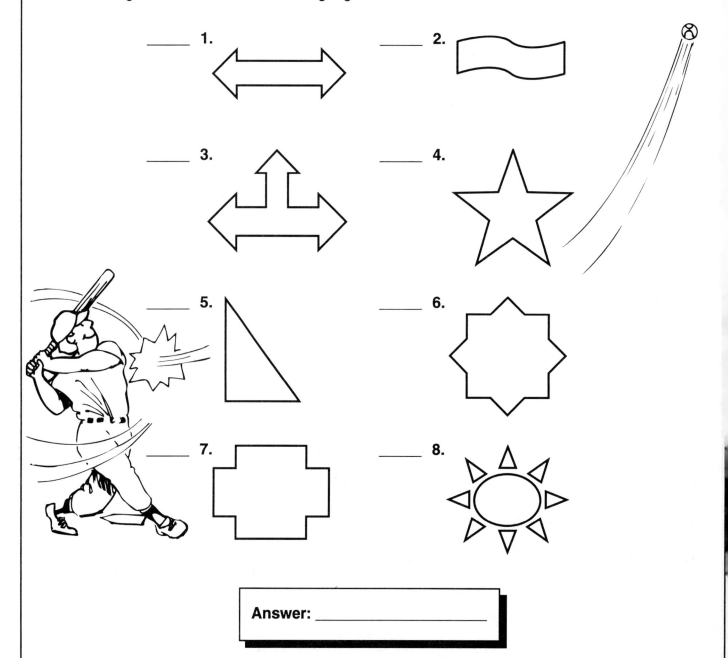

_____ 1.

_____ 2.

_____ 3.

_____ 4.

_____ 5.

_____ 6.

_____ 7.

_____ 8.

Answer: _____

FS122011 Introduction to Geometry Made Simple • © Frank Schaffer Publications, Inc.

Two of a Kind

Given polygon ABCDE ≅ polygon FGHIJ, complete the questions below.

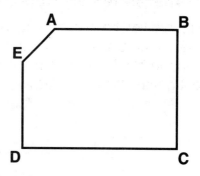

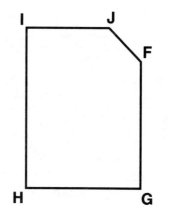

1. $\overline{AB} \cong$? _____

2. $\overline{DC} \cong$? _____

3. $\angle E \cong \angle$? _____

4. $\angle C \cong \angle$? _____

5. The length of $\overline{BC}$ is 12 cm. What is the length of $\overline{GH}$? _____

6. The length of $\overline{AE}$ is 4 cm. What is the length of $\overline{FJ}$? _____

7. $m\angle E = 135°$. What is the $m\angle J$? _____

8. $m\angle A = 150°$. What is the $m\angle F$? _____

9. $\overline{AB} + \overline{CD} + \overline{EA} \cong$? _____

10. $m\angle F + m\angle H + m\angle I \cong$? _____

Jelly Jar Joy

What is a jelly jar's favorite month of the year?

To find out, calculate the value of a, b, and c in each problem, given that each pair of polygons is congruent. Write your answers in the blanks. Find each of the values of b at the bottom of the page. Write the corresponding letter of its problem above the value to spell out the answer to this jelly riddle.

M.

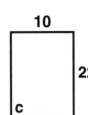 

a = _____
b = _____
c = _____

U.

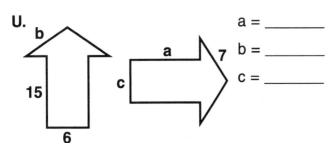

a = _____
b = _____
c = _____

A.

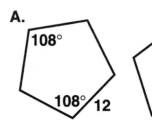

a = _____
b = _____
c = _____

J.

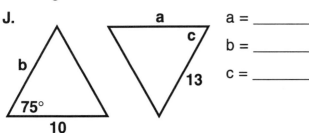

a = _____
b = _____
c = _____

Y.

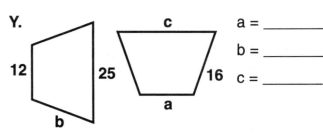

a = _____
b = _____
c = _____

R.

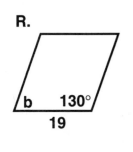 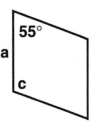 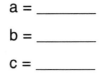

a = _____
b = _____
c = _____

A.

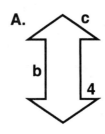 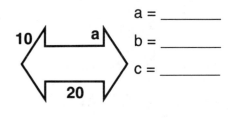

a = _____
b = _____
c = _____

___ ___ ___   -   ___ ___ ___ ___
13   12   22   7   20   55° 16

   FS122011 Introduction to Geometry Made Simple ▪ © Frank Schaffer Publications, Inc.

Monumental Mountains

What is the highest mountain in the world, reaching 29,029 feet high?

To find out, use your knowledge of the sum of the interior angles of polygons to find the measure of each angle A below. Shade in the boxes that contain your angle measures. Read across the remaining unshaded boxes to spell out the answer to this mountainous question.

1.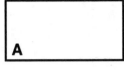

2.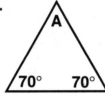

3.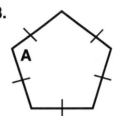

4.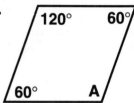

5.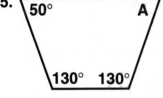

6.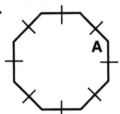

7.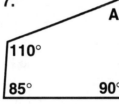

8.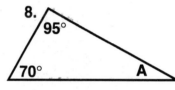

M	O	G	Y	U
95°	110°	75°	90°	45°
N	H	T	E	R
125°	108°	10°	25°	50°
V	E	B	R	J
140°	100°	120°	130°	135°
E	D	S	M	T
55°	15°	80°	40°	85°

Answer: _____

Triangles and Quadrilaterals

Every math student will greatly benefit from the activities involving triangles and quadrilaterals in this section. Allow students ample opportunity to work with manipulatives and time to complete several examples with your guidance. Be sure students gain a conceptual understanding of the concepts to the right before proceeding through the independent student activity pages (pages 34–42).

Present everyday situations to students in which they may use their new skills. For example, students can use their knowledge of triangles and quadrilaterals when constructing buildings or objects and when carpeting or tiling an area. Help students observe the world in which they live and identify their own connections involving working with triangles and quadrilaterals.

CONCEPTS

The ideas and activities presented in this section will help students explore the following concepts:

- finding angle measures in triangles
- classifying triangles by their sides
- classifying triangles by their angles
- identifying congruent triangles
- drawing triangles
- constructing triangles
- using properties of quadrilaterals

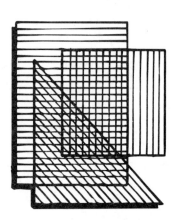

GETTING STARTED

Ask students to answer and explain the questions below.

- How do the areas of triangles and quadrilaterals relate to one another?
- How are a square and a rectangle similar? How do they differ?
- How are a square and a rhombus similar? How do they differ?
- How are a parallelogram and a rectangle similar? How do they differ?
- Show the relationship of a square, rectangle, rhombus, and parallelogram to each other. What properties do all of them share?

TRIANGLE TRIVIA

Ask students to state whether each statement below about triangles is true or false. Be sure they explain and justify each answer.

- An isosceles triangle can also be an equilateral triangle. (F—equilateral has three congruent sides)
- A scalene triangle can also be an isosceles triangle. (F—scalene has no sides congruent)
- An equilateral triangle can also be an isosceles triangle. (T)
- A right triangle can be equiangular. (F—equiangular has three angles congruent)
- An obtuse triangle can be a right triangle. (F—obtuse is greater than 90°)
- An acute triangle can be equiangular. (T)
- An equiangular triangle can also be obtuse. (F—equiangular must have three 60° angles)

FS122011 Introduction to Geometry Made Simple ▪ © Frank Schaffer Publications, Inc.

INVESTIGATING QUADRILATERALS

Divide students into pairs. Ask them to write, in their own words, a definition for a square, a rhombus, and a rectangle. Then have them draw several parallelograms on graph paper and measure each of the angles using a protractor. What do they find is true about the angles? (opposite angles are equal) Have them write a definition of a parallelogram using this information.

QUICK QUIZ

Ask students to write "yes" or "no" for each statement below. Be sure they explain their answers in detail.

- Is a square a rectangle? (yes)

- Is a rectangle a square? (no)

- Is a square a rhombus? (yes)

- Is a rhombus a square? (no)

- Is a parallelogram a quadrilateral? (yes)

- Are squares, rectangles, and rhombi all parallelograms? (yes)

TOYING WITH TRIANGLES

Ask students to draw several different triangles. Have them use scissors to cut out each triangle. Then have them cut off each of the three angles in each triangle. For each triangle they drew, ask them to draw a straight line with one point on it. Have them arrange the three cutout angles of each triangle on its own line around the one point. What do students observe about each set of three angles? What do they find the three angles' sum to equal? Based on the information students found, have them make a conjecture about the three angles of a triangle. (Three angles of a triangle will add up to 180°.)

QUICK APPLICATIONS

Ask students to solve the problems below based on their knowledge of triangles and quadrilaterals.

- One acute angle of a right triangle is 4 times as large as the other. What is the measure of each angle in the triangle? (18°, 72°, 90°)

- Could the sides of a right triangle be 2 inches, 4 inches, and 7 inches? (no)

- Draw a Venn diagram that shows the relationship of squares, rectangles, rhombi, and parallelograms to each other. Remember, they are all quadrilaterals!

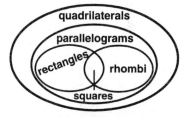

- Draw a Venn diagram that shows the relationship of right triangles, isosceles triangles, and equilateral triangles to each other. Remember, they are all triangles!

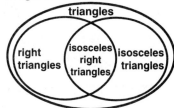

- Twenty-five equiangular triangles are placed in a row, creating a parallelogram. Each has 2-inch long sides. What is the distance around the row of triangles? (54 in.)

Group Activity

Divide students into groups. Ask them to solve the problem below with their group. They must explain their answers in detail.

One hundred rectangular tables, each with lengths of 8 feet and widths of 4 feet, are arranged side by side in a row. Two people can sit at each open side of a table. How many people can be seated at a row of 100 tables? (404 people) How many people can be seated at a row of 150 tables? (604 people)

FS122011 Introduction to Geometry Made Simple ▪ © Frank Schaffer Publications, Inc.

Law-Breaker

What happens when words misbehave and break the rules?

To find out, find the missing measure in each of the triangles below. Write the problem number in front of the corresponding answer. To spell out the answer at the bottom of the page, refer to the table and write the code letter that corresponds to the problem number given.

1.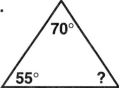

2.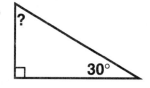

3.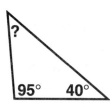

4.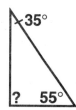

5.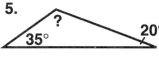

6.

7.

8.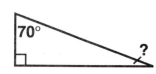

9.

Code Letter	Problem #	Answer
C		125°
D		55°
E		20°
G		90°
H		10°
N		60°
S		155°
T		45°
Y		50°

___ ___ ___ ___   ___ ___ ___   ___ ___ ___ ___ ___ ___ ___ ___ ___ **!**
3 9 8 7   4 8 3   6 8 2 3 8 2 5 8 1

FS122011 Introduction to Geometry Made Simple ▪ © Frank Schaffer Publications, Inc.

Singing Success

What is the name of the musical group that is considered to be the most successful band of all time—with the greatest record sales of any other group now and then? (Hint: "The _____")

To find out, classify each triangle below according to the measures of its sides. Circle the letter that represents the correct name of each triangle. Unscramble the circled letters to spell out the name of this record-making band.

1. 3 cm, 7 cm, 9 cm   (T) scalene   (S) equilateral   (U) isosceles

2. 11 in, 11 in, 11 in   (P) scalene   (B) equilateral   (L) isosceles

3. 12 m, 8 m, 12 m   (T) scalene   (I) equilateral   (S) isosceles

4. 2 ft, 4 ft, 6 ft   (E) scalene   (C) equilateral   (I) isosceles

5. 5 m, 10 m, 5 m   (G) scalene   (M) equilateral   (L) isosceles

6. 3 cm, 3 cm, 3 cm   (E) scalene   (A) equilateral   (I) isosceles

7. 7 in, 7.5 in, 8 in   (E) scalene   (R) equilateral   (T) isosceles

Answer: _____

Heavy Hail

The heaviest hailstones on record were found in Bangladesh on April 14, 1986, killing 92 people. These hailstones weighed a little over how many pounds?

To find out, classify each triangle below by its angles. Write your answer next to each problem number. Count the number of triangles that are obtuse. The total will give you the answer to this unbelievable record weight.

1. 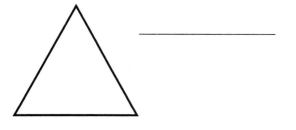 _____

5. 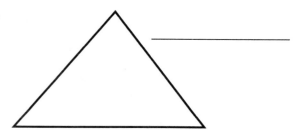 _____

2. 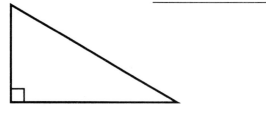 _____

6. 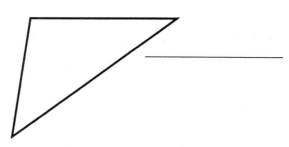 _____

3. 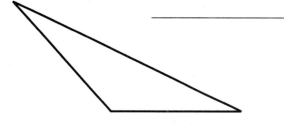 _____

7.  _____

4. _____

8. 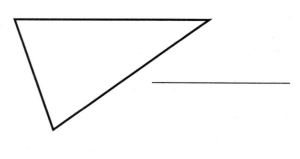 _____

Answer: _____

FS122011 Introduction to Geometry Made Simple ▪ © Frank Schaffer Publications, Inc.

Puzzlin' Through Angles

Answer each question using the figure below.

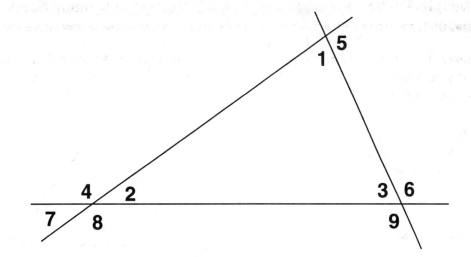

1. What two measures have the sum equal to the m∠5? _____

2. What two measures have the sum equal to the m∠9? _____

3. What two measures have the sum equal to the m∠4? _____

Given m∠1 = 80° and m∠2 = 35°, find the measure of each angle in the problems below.

4. m∠3 _____

5. m∠4 _____

6. m∠5 _____

7. m∠6 _____

8. m∠7 _____

9. m∠8 _____

10. m∠9 _____

Identifying angle measures **37**

Swimming Through Triangles

What household appliance will never be able to swim?

To find out, identify whether each pair of triangles is congruent by SSS, SAS, or ASA. Circle the letter that represents this characteristic. Place the circled letters in the blanks at the bottom of the page above the corresponding problem numbers.

1. 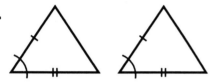   (O) SSS   (T) SAS   (L) ASA

2. 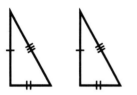   (H) SSS   (K) SAS   (M) ASA

3. 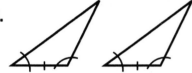   (Y) SSS   (B) SAS   (E) ASA

4. 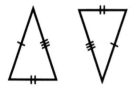   (S) SSS   (A) SAS   (C) ASA

5.    (J) SSS   (I) SAS   (T) ASA

6.    (W) SSS   (R) SAS   (N) ASA

7. 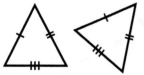   (K) SSS   (M) SAS   (F) ASA

___ ___ ___   ___ ___ ___ ___
 1   2   3   4   5   6   7

FS122011 Introduction to Geometry Made Simple • © Frank Schaffer Publications, Inc.

Toying With Triangles

Draw and label each triangle below.

1. ΔEFG is acute and equilateral.

2. ΔABC is obtuse and scalene.

3. ΔXYZ is right and isosceles.

4. ΔMNO is acute and scalene.

5. ΔJKL is obtuse and isosceles.

6. ΔRST is right and scalene.

Rules of Construction

Use a compass and a straightedge to construct the triangles below using each indicated rule.

1. triangle MNO congruent to triangle RST using the ASA rule

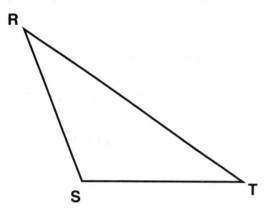

2. triangle XYZ congruent to triangle HIJ using the SSS rule

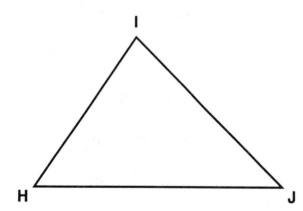

3. triangle ABC congruent to triangle PQR using the SAS rule

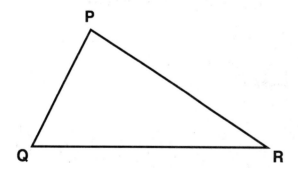

FS122011 Introduction to Geometry Made Simple ▪ © Frank Schaffer Publications, Inc.

Private Property

Can you possibly name all of the properties of each of the quadrilaterals listed below? To help you get started, find the letters on the right that represent the properties of each shape. Write the letters in the blank below each problem number. Remember, the properties are going to be used more than once.

1. parallelogram

_____

2. trapezoid

_____

3. rectangle

_____

4. square

_____

5. rhombus

_____

A. a quadrilateral

B. one pair of parallel sides

C. opposite sides parallel

D. four sides congruent

E. opposite sides congruent

F. four angles congruent

G. opposite angles congruent

H. four right angles

I. two lines of symmetry

J. no lines of symmetry

K. four lines of symmetry

L. one line of symmetry

Perfect Numbers

A number is said to be a perfect number if it is equal to the sum of its divisors other than itself. What is the lowest perfect number?

To find out, state whether each statement below is true or false. Count the number of your false answers. Your total will be the answer to the perfect number question.

_____ **1.** A quadrilateral is a square.

_____ **2.** A parallelogram is a quadrilateral.

_____ **3.** A square is a rhombus.

_____ **4.** All squares and rectangles are parallelograms.

_____ **5.** A trapezoid is a parallelogram.

_____ **6.** A rhombus is a rectangle.

_____ **7.** A rectangle is a square.

_____ **8.** A trapezoid is a quadrilateral.

_____ **9.** A rhombus is a square.

_____ **10.** All quadrilaterals are parallelograms.

Answer: _____

FS122011 Introduction to Geometry Made Simple ▪ © Frank Schaffer Publications, Inc.

Perimeter and Circumference

Every math student will greatly benefit from the activities involving perimeter and circumference in this section. Allow students ample opportunity to work with manipulatives and time to complete several examples with your guidance. Be sure students gain a conceptual understanding of the concepts below before proceeding through the independent student activity pages (pages 45–49).

Present everyday situations to students in which they may use their new skills. For example, students can use their knowledge of perimeter and circumference when working with revolutions of a circular object and when building a fence or a border of some kind. Help students observe the world in which they live and identify their own connections involving working with perimeter and circumference.

CONCEPTS

The ideas and activities presented in this section will help students explore the following concepts:
- perimeter of polygons
- perimeter of squares, rectangles, and triangles
- circumference of circles

PERIMETER PARTICULARS

Ask students to find the perimeter of each object below. Have them name two different methods they can use to find the perimeter of each. Ask students to explain in detail which method they think is easier and less time-consuming and why they think this.

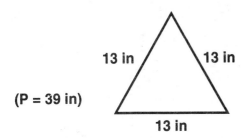

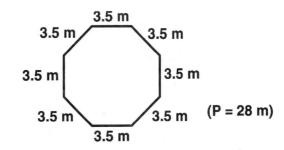

QUICK QUIZ

Ask students to answer the questions below about the circumference of circles.

- Can you estimate the circumference of a circle when all you have is the dimension of the diameter? (yes) If so, explain how.

- Can you estimate the circumference of a circle when all you have is the dimension of the radius? (yes) If so, explain how.

- Describe the relationship between the circumference of a circle and the perimeter of a polygon.

CIRCUMFERENCE INVESTIGATION

Ask students to compare the circumference of a circle with a diameter of 6 to the circumference of a circle with a radius of 3. (They are the same.) Ask them what they find and why they think this can be so. Ask them what must be true about circles and their diameters and radii for this outcome to always occur. Have them give at least three other examples that yield the same outcome.

PERIMETER AND CIRCUMFERENCE IN COMPARISON

Ask students to compare the perimeters of the square and the circle below and answer each question.

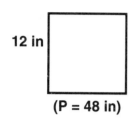 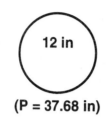

12 in

(P = 48 in)

12 in

(P = 37.68 in)

- Which perimeter did you find to be the greatest? (square)

- Why do you think this is true?

- Will the perimeter of a square always be greater than that of a circle when the diameter and side of a square are the same? (yes) Why or why not?

PRACTICE AND APPLY

Ask students to work the problems below and provide explanations with their answers.

- How will the circumference change if you double the diameter of a circle? Give two examples to help demonstrate your answer. (It will double the circumference.)

- If you want to make your own basketball hoop, what length of metal bar will you need if the diameter of the hoop is 50 centimeters? (157 cm)

- What is the diameter of a 30-inch bicycle wheel? (9.6 in) How many revolutions will this wheel make in a 20-mile bicycle race? (1,267,000 ÷ 9.6 = 132,000)

- Suppose the circumference of a circular pool is 50.24 feet. How many feet would you swim if you swam across the pool, swimming directly through the center? (16 ft)

Pi Project

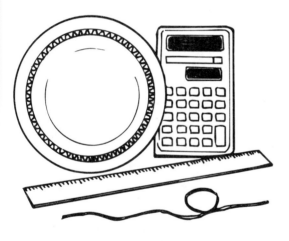

Divide students into groups of three or four. Each group will need a pencil and paper, four circular objects of different sizes, a piece of string, a straightedge, and a calculator. Ask the groups to measure the circumference and the diameter of each object using the string. Then, having both of these measurements for each circular object, ask students to divide each circumference by each diameter. Have them compare their results with each of the other groups and then answer the questions below, providing ample explanations.

- After each group compared the results with every other group, what is the average value of π the class discovered as a whole?

- Does each of your calculators have a π key? If so, what is the value of π on your calculator?

- How does the average value of π the class discovered compare to that of the value of π on the calculator?

- Explain the process that was just used in calculating the value of π using the diameter and circumference of each circular object. What are some of the errors that could have affected the calculation of this value to make it somewhat inaccurate compared to that of the actual value of π?

Powerful Perimeter

What is the name of the superhero who drinks apple juice and scales tall buildings?

To find out, find the perimeter of each polygon below. Circle the letter that represents each answer. Unscramble the circled letters to spell out the name of this amazing superhero.

1.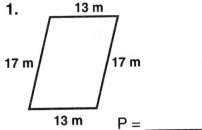

P = _____

2.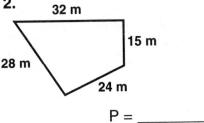

P = _____

3.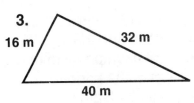

P = _____

4.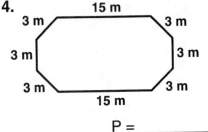

P = _____

5.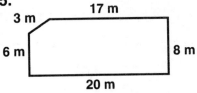

P = _____

6.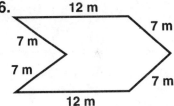

P = _____

7.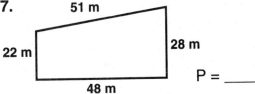

P = _____

8.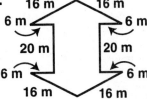

P = _____

(L) 42 m	(A) 60 m	(H) 50 m
(C) 48 m	(K) 32 m	(E) 99 m
(M) 52 m	(D) 128 m	(I) 54 m
(O) 67 m	(N) 88 m	(R) 149 m

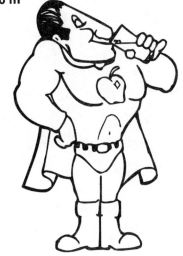

Answer: _____

Name____

Excellent Education

Dan and Helen Fagan of Bessemer, Alabama, have seen their children receive bachelor's degrees from a university or college between the years 1969 and 1992. This family holds the world record for the most graduates in a family. How many children graduated in this family?

To find out, find the perimeter of each square, rectangle, and triangle. Write the corresponding letter of each problem above its perimeter at the bottom of the page to spell out the answer to this educational feat.

N. l = 9 in, w = 12 in P = ________

E. l = 11 in, w = 20 in P = ________

F. l = 12.5 in, w = 10.5 in P = ________

T. s = 12 in

P = ________

I. s = 9 in

P = ________

E. a = 14 in, b = 15 in, c = 16 in

P = ________

F. a = 21 in, b = 4 in, c = 8 in

P = ________

___	___	___	___	___	___	___
33 in	36 in	46 in	48 in	45 in	62 in	42 in

Tree Knowledge

What kind of notebook grows on a tree?

To find out, use 3.14 for π to find the circumference of each circle below. Shade in the boxes that contain your answers. Read across the remaining unshaded boxes to spell out the answer to this tree riddle.

1. C = _____
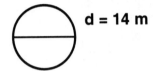
d = 14 m

2. C = _____

r = 4 m

3. C = _____
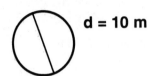
d = 10 m

4. C = _____
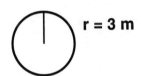
r = 3 m

5. C = _____
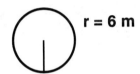
r = 6 m

6. C = _____
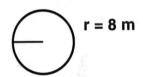
r = 8 m

7. C = _____
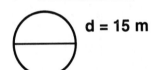
d = 15 m

8. C = _____
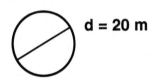
d = 20 m

9. C = _____
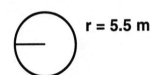
r = 5.5 m

L	H	T	O	M	O
37.78 m	47.1 m	25.12 m	35.54 m	18.84 m	51.24 m
S	**G**	**E**	**N**	**L**	**J**
44.96 m	37.68 m	32.4 m	50.24 m	63.8 m	43.96 m
E	**V**	**B**	**A**	**T**	**F**
26.12 m	34.54 m	31.4 m	18.85 m	62.8 m	47.5 m

Answer: _____

Baseball Blow-Out

This state hosted the first professional baseball game ever to be played in the world in 1846. What is the name of this state?

To find out, use 3.14 for π to find the circumference of each circle below. Round your answers to the nearest tenth. Circle the letter that represents each circumference. Place the circled letters in the blanks at the bottom of the page above the corresponding problem numbers.

1. d = 16 m   (I) 51.0 m   (N) 50.2 m   (U) 52.2 m

2. r = 3.5 m   (E) 22.0 m   (L) 21.9 m   (Y) 20.3 m

3. d = 9 m   (H) 26.4 m   (W) 28.3 m   (L) 28.4 m

4. r = 8.5 m   (I) 54.0 m   (T) 53.9 m   (J) 53.4 m

5. d = 21 m   (E) 65.9 m   (N) 66.0 m   (F) 65.8 m

6. r = 9.5 m   (B) 59.8 m   (O) 60.0 m   (R) 59.7 m

7. d = 32 m   (D) 100.0 m   (S) 100.5 m   (I) 100.4 m

8. r = 4.3 m   (E) 27.0 m   (P) 27.1 m   (S) 26.9 m

9. d = 13 m   (M) 40.9 m   (Y) 40.8 m   (E) 41.0 m

___ ___ ___   ___ ___ ___ ___ ___ ___
 1   2   3   4   5   6   7   8   9

FS122011 Introduction to Geometry Made Simple ▪ © Frank Schaffer Publications, Inc.

Skating Through Circumference

Use 3.14 for π to find the missing measure in each of the circular objects described below. Sketch each picture and write the answer in the blank.

1. the diameter of a supreme pizza from "Mama Mario's" with a circumference of 50.24 inches

   d = _____

2. the radius of a Ferris wheel at the school circus with a circumference of 219.8 feet

   r = _____

3. the diameter of a soccer ball with a circumference of 31.4 inches

   d = _____

4. the radius of a scoop of double chocolate fudge supreme ice cream with a circumference of 34.54 centimeters

   r = _____

5. the diameter of a circular pool with a circumference of 78.5 feet

   d = _____

6. the radius of a basketball hoop with a circumference of 43.96 inches

   r = _____

7. the diameter of the earth with a circumference of 24,887.64 miles

   d = _____

8. the radius of a hamburger with a circumference of 12.56 inches

   r = _____

Area of Polygons

Every math student will greatly benefit from the activities involving the area of polygons in this section. Allow students ample opportunity to work with manipulatives and time to complete several examples with your guidance. Be sure students gain a conceptual understanding of the concepts below before proceeding through the independent student activity pages (pages 52–59).

Present everyday situations to students in which they may use their new skills. For example, students can use their knowledge of the area of polygons when calculating the amount of space a house will occupy in a yard and when calculating the area in which a sprinkler will spray. Help students observe the world in which they live and identify their own connections involving working with the area of polygons.

CONCEPTS

The ideas and activities presented in this section will help students explore the following concepts:

- area of triangles
- area of parallelograms, rectangles, and squares
- area of shaded regions
- area of trapezoids
- area of circles

GETTING STARTED

Ask students to discover the formula for the area of a circle using the shape of a parallelogram by guiding them through the steps below.

- Draw any size circle on a piece of unlined paper.
- Mark the center of the circle.
- Draw lines through the center of the circle, separating the circle into eight congruent sections.
- Cut each section out and fit the sections together to create a parallelogram.
- Knowing that $C = \pi r$, what is the base of the parallelogram?
- What is the height of the parallelogram?
- Putting the two previous values together and using the knowledge that the area of a parallelogram is the base times the height, what did you formulate for the area of a circle? Explain your reasoning.

QUICK QUESTION

Ask students to explain how the formula for the area of a parallelogram can help them to determine the formula for the area of a triangle. Have them give two examples to provide concrete proof that this relationship will always be true.

FS122011 Introduction to Geometry Made Simple • © Frank Schaffer Publications, Inc.

TRIANGLES TO TRAPEZOIDS

Ask students to help formulate the formula for the area of a trapezoid by answering the questions below. You will want to draw the figure to the right for them to refer to.

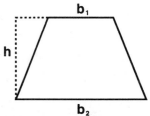

Given that the diagonal separates this trapezoid into two triangles, what is the height of the two triangles? (h)

What is the formula for the area of each of the triangles? ($A = \frac{1}{2}hb$)

- If you add the two formulas together, what is the result? ($\frac{1}{2}h \cdot b_1 + \frac{1}{2}h \cdot b_2$)

- Use the distributive property to simplify the previous result. The result will be the formula for the area of a trapezoid. Explain the process that created this outcome. [$\frac{1}{2}h(b_1 + b_2)$]

RECTANGLE CREATION

Ask students to follow the directions below and answer each of the questions.

- Draw a parallelogram with a height of 5 cm and a base of 8 cm.

- Draw a dotted line showing the height of the parallelogram.

- Cut a triangular section from the parallelogram, cutting along the dotted line.

- Place the triangular section at the opposite side of the parallelogram. What shape does this create?

- If a parallelogram and a rectangle have equal bases and equal heights, how will their areas relate? Explain your reasoning.

NUMBER CRAZE

Ask students to use the drawing to solve the problems below. Tell them that the perimeter of the square inside the parallelogram is 24 inches.

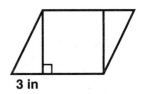

3 in

- Find the area of the square. ($36\ \text{in}^2$)

- Find the area of the parallelogram. ($54\ \text{in}^2$)

- Can you find the perimeter of the parallelogram? (no) Why or why not?

Circular Quiz

Ask students to answer the questions below about circles and provide ample explanations.

- What does the r^2 mean in the formula for the area of a circle? (radius x radius)

- If you only know the diameter of a circle, can you still find its area? (yes) If so, explain how.

- What is the difference between the area of a circle and the circumference of a circle? (*Circumference* is the distance around, and *area* is the amount of surface inside.)

- The world's largest chocolate chip cookie had an area of 5241.5 square feet. If its diameter had been increased by 4 feet, what would have been the radius? Round your answer to the nearest foot. (r = 43 feet)

Name_____

Triangle Trivia

Find the area of each triangle described below. Sketch each triangle and label the base and height.

_____ **1.** a triangular pool with a base of 35 feet and a height of 50 feet

_____ **2.** a triangular piece of stained glass with a base of 13 inches and a height of 25 inches

_____ **3.** a triangular kite with a base of 60 centimeters and a height of 95 centimeters

_____ **4.** one side of a triangular roof with a base of 75 feet and a height of 40 feet

_____ **5.** a triangular diamond with a base of 15 millimeters and a height of 3.5 millimeters

_____ **6.** a triangular pool table with a base of 9 feet and a height of 15 feet

_____ **7.** a triangular piece of gold with a base of 9.8 millimeters and a height of 10.1 millimeters

_____ **8.** a triangular piece of art with a base of 30 inches and a height of 48 inches

FS122011 Introduction to Geometry Made Simple ▪ © Frank Schaffer Publications, Inc.

Food Fill

What state grows more food than any other state in the United States?

To find out, find the area of each parallelogram, rectangle, and square below. Write the corresponding letter of each shape above its area at the bottom of the page to spell out the answer to this fun food fact.

N. 13 m A = _____

I. A = _____ 70 m 25 m

L. A = _____ 6 m 22 m

R. 40 m 32 m A = _____

A. 13 m 15 m A = _____

F. 16 m A = _____

O. 11 m A = _____

A. 19 m 33 m A = _____

C. A = _____ 2.5 m 12 m

I. A = _____ 8 m 1.5 m

_____ _____ _____ _____ _____ _____ _____ _____ _____ _____
30 m² 195 m² 132 m² 12 m² 256 m² 121 m² 1280 m² 169 m² 1750 m² 627 m²

Workin' Out in Math

What does a brontosaurus get if it works too many math problems at once?

To find out, find the area of each shaded region below. Circle the letter that represents each area. Place the circled letters in the blanks at the bottom of the page above the corresponding problem numbers.

1.

(D) 65   (S) 75   (I) 60

2.

(U) 270   (I) 276   (G) 200

3.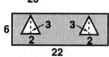

(N) 126   (P) 120   (Y) 125

4.

(A) 165   (L) 170.5   (O) 164.5

5.

(S) 900   (T) 980   (H) 908

6.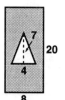

(E) 150   (O) 146   (K) 140

7.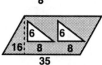

(A) 510   (R) 512   (O) 525

8.

(O) 632   (C) 600   (E) 636

____ ____ ____ ____ - ____ ____ ____ ____
  1   2   3   4   5   6   7   8

FS122011 Introduction to Geometry Made Simple ▪ © Frank Schaffer Publications, Inc.

Great Wall of Shade

The Great Wall of China is known to be the longest wall in the world. It is 15–39 feet high and up to 32 feet thick. How many miles long is this Great Wall?

To find out, find the area of each shaded region below. Write each answer in the blank. Add these shaded areas together. The total will be the number of miles this wall covers.

1.

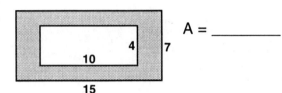

A = _____

2.

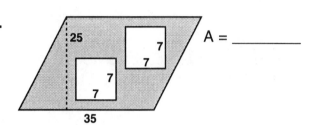

A = _____

3.

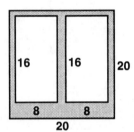

A = _____

4.

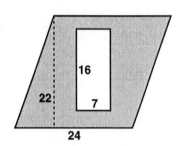

A = _____

5.

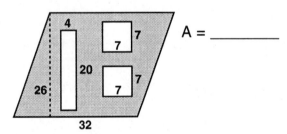

A = _____

6.

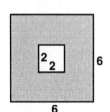

A = _____

7.

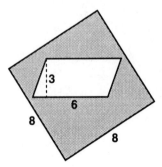

A = _____

8.

A = _____

Answer: _____

Tangled Up in Trapezoids

Find the area of each trapezoid described below. Sketch each picture and label the bases and height.

1. a trapezoid with bases of 12 m and 14 m and a height of 15 m

   A = _____

2. a trapezoid with bases of 10 mm and 15 mm and a height of 20 mm

   A = _____

3. a trapezoid with bases of 8 ft and 6 ft and a height of 16 ft

   A = _____

4. a trapezoid with bases of 20 cm and 30 cm and a height of 45 cm

   A = _____

5. a trapezoid with bases of 18 in and 26 in and a height of 36 in

   A = _____

6. a trapezoid with bases of 35 yd and 45 yd and a height of 50 yd

   A = _____

FS122011 Introduction to Geometry Made Simple • © Frank Schaffer Publications, Inc.

Polly's Pizza Parlor

Find the area of each of Polly's special pizzas below. Use 3.14 for π. Sketch each picture and label the given radius or diameter.

1. a medium taco pizza with extra cheese and a radius of 7 inches

   A = _____

2. a small personal pizza with pineapple, bacon, anchovies, and tomatoes and a diameter of 6 inches

   A = _____

3. a large family pizza with "Polly's Pleasers" and a radius of 9 inches

   A = _____

4. an extra large pizza with barbecue chicken, onions, and Polly's famous sauce and a diameter of 24 inches

   A = _____

5. You and your teammates went out for pizza after the big game. You ordered 3 Super Size pizzas with "the works," each with a radius of 13 inches. What was the total area of pizza you and your team ate?

   A = _____

6. You and your friends had a pizza party. You were only able to eat half of the "Power Pizza" that you ordered. If this pizza had a diameter of 28 inches, what was the area of the pizza that was left?

   A = _____

Bubblin' Through Bathroom Area

What do lizards use to cover their bathroom walls?

To find out, find the area of each figure below. Shade in the boxes that contain your answers. Read across the remaining unshaded boxes to spell out the answer to this bathroom riddle. Use π = 3.14.

1.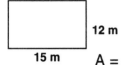
12 m
15 m
A = _____

2.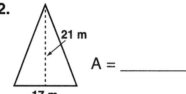
21 m
17 m
A = _____

3.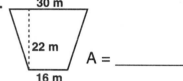
30 m
22 m
16 m
A = _____

4.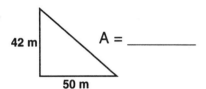
42 m
50 m
A = _____

5.
9 m
12 m
A = _____

6.
11 m
A = _____

7.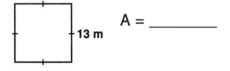
13 m
A = _____

8.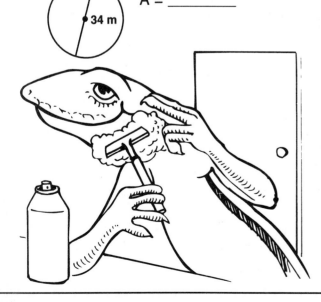
34 m
A = _____

H	R	Y	E
1050 m²	350 m²	169 m²	510 m²
P	**G**	**T**	**O**
110 m²	379.94 m²	380.76 m²	180 m²
U	**I**	**R**	**L**
506 m²	908.54 m²	108 m²	175 m²
E	**K**	**S**	**Y**
360 m²	907.46 m²	502 m²	178.5 m²

Answer: _____

FS122011 Introduction to Geometry Made Simple ▪ © Frank Schaffer Publications, Inc.

Billions of Area

This man is known to be the youngest self-made billionaire in the world. He was 20 when he set up his company, Microsoft of Seattle, Washington, and became a billionaire at the age of 31. What is this billionaire's name?

To find out, find the area of each shaded region below. Circle the letter that represents each answer. Unscramble the circled letters to spell out the first and last name of this amazing young entrepreneur. Use $\pi = 3.14$.

1.

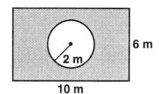

10 m
6 m
2 m

A = _____

2.

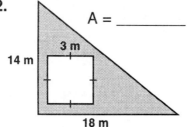

A = _____
14 m
3 m
18 m

3.

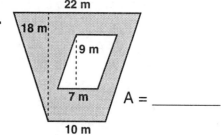

22 m
18 m
9 m
7 m
10 m

A = _____

4.

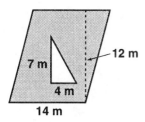

7 m
12 m
4 m
14 m

A = _____

5.

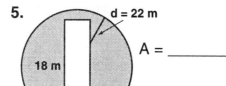

d = 22 m
18 m
5 m

A = _____

6.

15 m
11 m
9 m
22 m

A = _____

7.

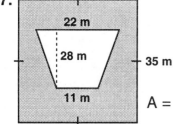

22 m
28 m
35 m
11 m

A = _____

8.

40 m
3 m
25 m
6 m
20 m
24 m

A = _____

9.

6 m
d = 26 m

A = _____

I. 117 m²	**L.** 763 m²	**G.** 154 m²	**H.** 50.2 m²	**B.** 536.74 m²	**W.** 422 m²
T. 417.62 m²	**O.** 245 m²	**S.** 115.5 m²	**L.** 47.44 m²	**E.** 289.94 m²	**A.** 225 m²

Answer: _____

Surface Area and Volume of Three-Dimensional Figures

Every math student will greatly benefit from the activities involving the surface area and volume of three-dimensional figures in this section. Allow students ample opportunity to work with manipulatives and time to complete several examples with your guidance. Be sure students gain a conceptual understanding of the concepts to the right before proceeding through the independent student activity pages (pages 62–74).

Present everyday situations to students in which they may use their new skills. For example, students can use their knowledge of the surface area and volume of three-dimensional figures when manufacturing containers for storage and when creating models of buildings and structures in business. Help students observe the world in which they live and identify their own connections involving working with the surface area and volume of three-dimensional figures.

CONCEPTS

The ideas and activities presented in this section will help students explore the following concepts:

- identification of pyramids, prisms, cylinders, cones, and spheres
- surface area of prisms
- surface area of pyramids
- surface area of cylinders
- surface area of cones
- surface area of spheres
- volume of prisms and pyramids
- volume of cylinders
- volume of cones and spheres

GETTING STARTED

Explain to students the idea of the prefix *poly* and what it means exactly. (many) Then have them explain how this relates to the geometric shapes about which they will be studying. Ask them to give several examples of the different polyhedrons that are contained within the classroom.

VERIFYING VOLUME

Divide students into groups of three. Provide each group with small cubes, each containing the same volume. Ask students to stack the small cubes together to create one larger size cube. Have them count the number of cubes they used to create the larger cube and state its volume. Ask students to verify the volume they found for the larger cube by using the formula $V = e^3$. Make sure they explain their results.

QUICK QUIZ

Ask students to write *cylinder*, *cone*, or *sphere* to describe the shape in each description below.

- Has two parallel surfaces (cylinder)
- Contains one flat surface and one curved surface (cone)
- No flat surface (sphere)
- Two flat surfaces and one curved surface (cylinder)
- Ideal shape for a basketball (sphere)
- Ideal shape for a can of soup (cylinder)
- Contains the same appearance any way you look at it (sphere)
- Shape you wouldn't see food containers made to look like (sphere or cone)

FS122011 Introduction to Geometry Made Simple ■ © Frank Schaffer Publications, Inc.

DESIGNATED DESCRIPTIONS

Ask students to describe each of the polyhedrons below. Have them include the shape of the base, or bases, the shape and number of the faces, and the number of vertices.

- hexagonal pyramid

- triangular prism

- octagonal prism

- pentagonal pyramid

shape of base(s)	shape and # of faces	# of vertices
hexagon	triangle, 6	7
triangle	rectangle, 3	6
octagon	rectangle, 8	16
pentagon	triangle, 5	6

THREE-DIMENSIONAL PARTS

Draw the sketches below on the board. Ask students to name the shape of each one, given the unfolded parts.

(cylinder)

(rectangular pyramid)

(rectangular prism)

(triangular prism)

(pentagonal pyramid)

FIGURING FORMULAS

Ask students to explain how the formulas $V = lwh$ and $V = Bh$ for finding the volume of a rectangular prism relate to one another. Ask them if it matters which formula they should use or if they will both consistently give the same result. Have them explain their answers in complete detail.

AMAZING APPLICATIONS

Write the questions below on the board and have students answer them in a notebook. Check them when students complete the work.

- Mr. Andrews is cleaning his glass paperweight, which is shaped like a pentagonal pyramid. How many surfaces does he have to clean? (6)

- Your friend's aquarium leaks. Therefore, he needs to seal the edges more properly. If the tank is the shape of a hexagonal prism, how many edges does your friend have to seal? (18)

- You and your family like to wrap your gifts together the day before the winter holiday. The person who comes up with the most unique wrapping wins a prize. You found a box to use that is shaped like a decagonal prism and you want to put a red bow at each vertex. How many bows do you need? (20)

Group Activity

Divide students into groups of three. Ask them to use models like those in the "Three-Dimensional Parts" above and construction paper to create a cube, a prism, a pyramid, a cylinder, and a cone. Have students name three ways in which the shapes are alike and three ways in which the shapes are different.

FS122011 Introduction to Geometry Made Simple • © Frank Schaffer Publications, Inc.

Table Time

Fill in the table below with the characteristics of each shape.

Shape	Sketch of Shape	Number of Faces	Number of Vertices	Number of Edges
triangular pyramid				
triangular prism				
rectangular pyramid				
rectangular prism				
pentagonal pyramid				
pentagonal prism				
hexagonal pyramid				
hexagonal prism				

FS122011 Introduction to Geometry Made Simple ▪ © Frank Schaffer Publications, Inc.

Touchdowns in Math

What is the most touchdowns ever to be scored in a single game by the same player in the National Football League?

To find out, identify the shapes below. Count the total number of prisms. Your total will be the answer to the football trivia question.

1. _____

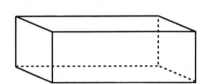

2. _____

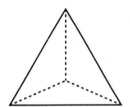

3. _____

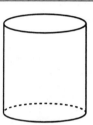

4. _____

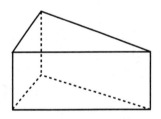

5. _____

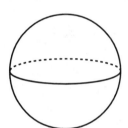

6. _____

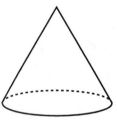

7. _____

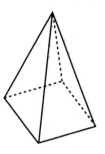

8. _____

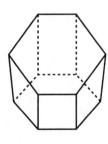

Answer: _____

Purrr-Fect Prisms

What is a cat's favorite thing to drink?

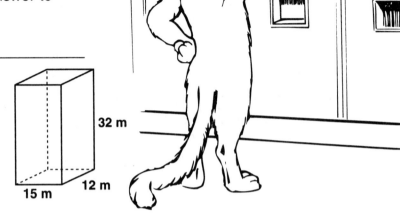

To find out, find the surface area of each prism below. Write the corresponding letter of each shape above its surface area at the bottom of the page to spell out the answer to this exciting cat riddle.

E. _____

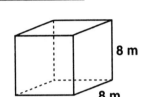
8 m
8 m
8 m

T. _____

32 m
15 m
12 m

A. _____

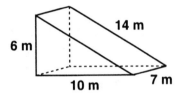
14 m
6 m
10 m
7 m

I. _____

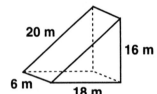
20 m
16 m
6 m
18 m

M. _____

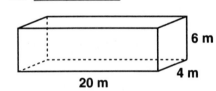
6 m
20 m
4 m

D. _____

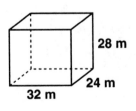
28 m
24 m
32 m

C. _____

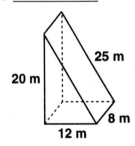
25 m
20 m
8 m
12 m

E. _____

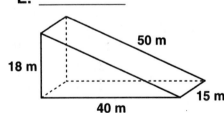
50 m
18 m
40 m
15 m

_____ _____ _____ _____ _____   _____ _____ _____
448 m² 612 m² 696 m² 384 m² 4672 m²   2088 m² 2340 m² 270 m²

FS122011 Introduction to Geometry Made Simple ▪ © Frank Schaffer Publications, Inc.

Puzzling Through Pyramids

Find the surface area of each pyramid below. Sketch each pyramid and label its dimensions.

_____ **1.** a pyramid with a base of 5 cm by 5 cm and an altitude of 10 cm

_____ **2.** a pyramid with a base of 7 in by 7 in and an altitude of 14 in

_____ **3.** a pyramid with a base of 12 m by 12 m and an altitude of 15 m

_____ **4.** a pyramid with a base of 16 ft by 16 ft and an altitude of 20 ft

_____ **5.** a pyramid with a base of 22 yd by 22 yd and an altitude of 25 yd

_____ **6.** a pyramid with a base of 30 mm by 30 mm and an altitude of 28 mm

_____ **7.** a pyramid with a base of 8.5 in by 8.5 in and an altitude of 13.5 in

_____ **8.** a pyramid with a base of 22.5 m by 22.5 m and an altitude of 32.5 m

Rainbow of Area

What is the only state in the United States to see a "moonbow"—a kind of rainbow you can see at night?

To find out, find the surface area of each cylinder below given the radius (or diameter) and the height. Circle the letter on the right that represents each answer. Unscramble these letters to find the answer to the interesting "moonbow" question. Use 3.14 for π.

1. radius = 4 m
   height = 10 m

2. diameter = 12 m
   height = 20 m

3. radius = 9 m
   height = 18 m

4. diameter = 16 m
   height = 25 m

5. radius = 3 m
   height = 9 m

6. diameter = 30 m
   height = 16 m

7. radius = 18 m
   height = 24 m

8. diameter = 34 m
   height = 22 m

(I) 350.2 m²   (T) 351.68 m²

(Y) 979.68 m²   (R) 2411.52 m²

(E) 1526.04 m²   (J) 1525.84 m²

(D) 4119.68 m²   (U) 1657.92 m²

(M) 227.03 m²   (K) 226.08 m²

(C) 2920.2 m²   (A) 2931.05 m²

(H) 4847.68 m²   (N) 4747.68 m²

(K) 4163.64 m²   (O) 4162.46 m²

Answer: _____

FS122011 Introduction to Geometry Made Simple ▪ © Frank Schaffer Publications, Inc.

Skipping Through Surface Area

Find the surface area of each shape below. Sketch each shape and label its dimensions. Use 3.14 for π.

_____ **1.** a rectangular prism with a length of 12 cm, a width of 9 cm, and a height of 15 cm

_____ **2.** a cylinder with a diameter of 22 in and a height of 32 in

_____ **3.** a pyramid with a square base of 6 ft by 6 ft and an altitude of 15 ft

_____ **4.** a rectangular prism with a length, width, and height of 12 m

_____ **5.** a cylinder with a radius of 30 in and a height of 35 in

_____ **6.** a pyramid with a square base of 11 mm by 11 mm and an altitude of 21 mm

_____ **7.** a rectangular prism with a length of 12 yd, a width of 15 yd, and a height of 18 yd

_____ **8.** a cylinder with a diameter of 25 in and a height of 30 in

Delicious Deliveries

Who delivers breakfast, lunch, and dinner and always completes his daily rounds?

To find out, find the surface area of each cone below. Use 3.14 for π. Shade in the boxes that contain your answers. Read across the remaining unshaded boxes to spell out the answer to this delightful riddle.

1. _____

15 m
6 m

2. _____

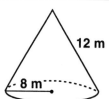
12 m
8 m

3. _____

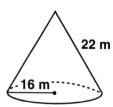
22 m
16 m

4. _____

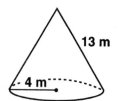
13 m
4 m

5. _____

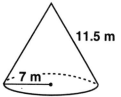
11.5 m
7 m

6. _____

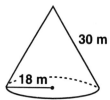
30 m
18 m

7. _____

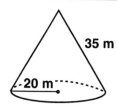
35 m
20 m

8. _____

40 m
15 m

T	U	H	I	E	P
1910.12 m²	2712.96 m²	3450 m²	1909.12 m²	503.5 m²	2590.5 m²
N	**M**	**G**	**E**	**K**	**A**
395.64 m²	214.62 m²	3456 m²	396.74 m²	502.4 m²	2710.86 m²
L	**F**	**M**	**A**	**J**	**N**
2690.4 m²	406.63 m²	210.45 m²	3290 m²	213.52 m²	398.87 m²

Answer: _____

FS122011 Introduction to Geometry Made Simple • © Frank Schaffer Publications, Inc.

Movie Mania

What is the most successful movie series of all time—grossing more than $1.776 billion?

To find out, find the surface area of each sphere below. Use 3.14 for π. Match each sphere in Column A to its surface area in Column B. Read down the column of written letters to discover the two-word answer to this record movie question.

COLUMN A

_____ 1.  r = 6 in

_____ 2.  d = 18 in

_____ 3.  r = 15 in

_____ 4.  d = 40 in

_____ 5.  r = 25 in

_____ 6.  d = 9 in

_____ 7. 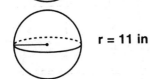 r = 11 in

_____ 8.  d = 56 in

COLUMN B

R. 1519.76 in²

N. 897.13 in²

S. 452.16 in²

B. 1900.01 in²

A. 2826 in²

Y. 342.17 in²

A. 254.34 in²

H. 113.04 in²

T. 1017.36 in²

U. 7824.23 in²

S. 9847.04 in²

O. 3456.89 in²

R. 5024 in²

J. 9820 in²

W. 7850 in²

Answer: _____

Coneheads vs. Ƒphereheads

Who won the contest for solving the most math problems correctly in an hour—the Coneheads or the Sphereheads?

To find out, find the surface area of each cone and sphere below. Add the total surface areas of the cones and the total surface areas of the spheres. The one that has the most surface area will be the winner of the math contest.

_____ **1.** a cone with a radius of 9 in and a slant height of 12 in

_____ **2.** a sphere with a diameter of 15 in

_____ **3.** a cone with a radius of 12 in and a slant height of 15 in

_____ **4.** a sphere with a radius of 11 in

_____ **5.** a cone with a radius of 15 in and a height of 20 in

_____ **6.** a sphere with a diameter of 30 in

_____ **7.** a cone with a radius of 8 in and a height of 22 in

_____ **8.** a sphere with a radius of 25 in

Score for Coneheads = _____

Score for Sphereheads = _____

Winner = _____

Clownin' Around

> **What did the English teacher say to the class clown?**

To find out, find the volume of each prism and pyramid below. Circle the letter that represents each area. Place each circled letter above its problem number at the bottom of the page to spell out the two-word answer to this fun riddle.

1. _____

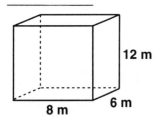

2. _____

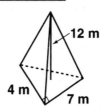

3. _____

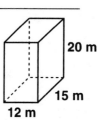

4. _____

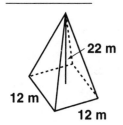

5. _____

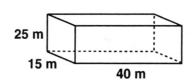

6. _____

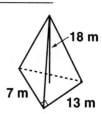

7. _____

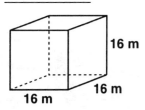

8. _____

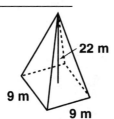

9. _____

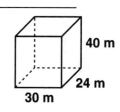

M. 1056 m³	**O.** 4096 m³	**C.** 576 m³	**H.** 581 m³
A. 15,000 m³	**N.** 28,800 m³	**G.** 32,400 m³	**O.** 56 m³
W. 594 m³	**U.** 856 m³	**M.** 3600 m³	**D.** 273 m³

___ ___ ___ ___ ___   ___ ___ ___ ___ **!**
 1   2   3   4   5   6   7   8   9

Coke™ Is It!

What state was the first place in the world to ever serve Coca-Cola™—in 1887?

To find out, find the volume of each cylinder below. Write the corresponding letter of each cylinder above its volume at the bottom of the page to spell out the answer to this interesting fact.

O. _____

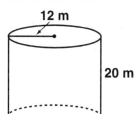

12 m

20 m

I. _____

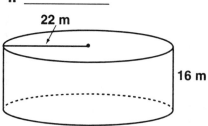

22 m

16 m

E. _____

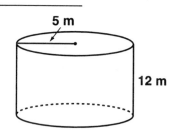

5 m

12 m

G. _____

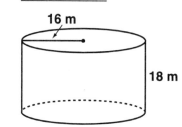

16 m

18 m

R. _____

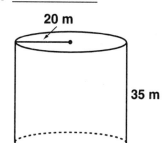

20 m

35 m

A. _____

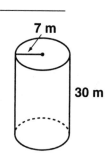

7 m

30 m

G. _____

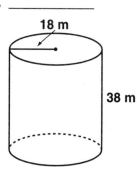

18 m

38 m

FIRST TO SERVE
Coke™

| 38,659.68 m³ | 942 m³ | 9043.2 m³ | 43,960 m³ | 14,469.12 m³ | 24,316.16 m³ | 4615.8 m³ |

FS122011 Introduction to Geometry Made Simple ▪ © Frank Schaffer Publications, Inc.

Name_____

Baby Boom

WAITING
ROOM

What is the highest number of babies reported to be born at a single birth?

To find out, find the volume of each cone and sphere below. Use 3.14 for π. Identify the problem number that contains the greatest volume. This will be the answer to the amazing baby fact.

1. _____

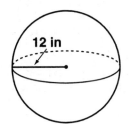

12 in

2. _____

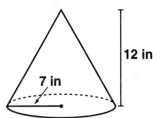

12 in

7 in

3. _____

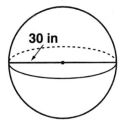

30 in

4. _____

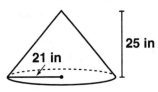

25 in

21 in

5. _____

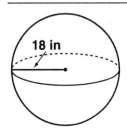

18 in

6. _____

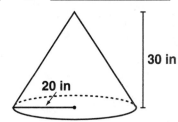

30 in

20 in

7. _____

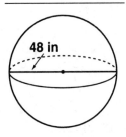

48 in

8. _____

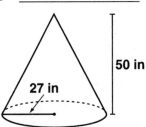

50 in

27 in

9. _____

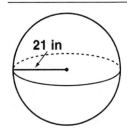

21 in

10. _____

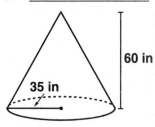

60 in

35 in

Answer: _____

Volume Verification

Find the volume of each shape described below. Use 3.14 for π. Sketch each shape and label its dimensions.

_____ **1.** a rectangular prism with a length of 14 in, a width of 9 in, and a height of 15 in

_____ **2.** a triangular pyramid with a height of 18 m and a triangle base with sides of 10 m and 12 m

_____ **3.** a cone with a radius of 20 mm and a height of 24 mm

_____ **4.** a cylinder with a diameter of 22 ft and a height of 16 ft

_____ **5.** a sphere with a radius of 18 cm

_____ **6.** a rectangular prism with a length of 20 m, a width of 30 m, and a height of 40 m

_____ **7.** a rectangular pyramid with a square base of 8 ft by 8 ft and a height of 21 ft

_____ **8.** a sphere with a diameter of 42 mm

_____ **9.** a cone with a radius of 24 in and a height of 32 in

_____ **10.** a cylinder with a radius of 25 cm and a height of 35 cm

 FS122011 Introduction to Geometry Made Simple ▪ © Frank Schaffer Publications, Inc.

Page 3

1. A
2. $\overline{BC}$, $\overline{CB}$
3. $\overrightarrow{DE}$
4. F
5. $\overleftrightarrow{GH}$, $\overleftrightarrow{HG}$
6. $\overrightarrow{JK}$
7. $\overleftrightarrow{MN}$, $\overleftrightarrow{NM}$
8. $\overline{PQ}$, $\overline{QP}$
9. $\overleftrightarrow{ST}$, $\overleftrightarrow{TS}$
10. $\overline{XY}$, $\overline{YX}$

Page 4

1.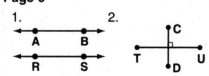
   P   Q
2. • Z
3. X   Y
4. A   B
5. R   M   S
6. U   C   D   T
7. G   O   H
8. J   L   K
9. V   R   E   T   W
10. X   B   Y

Page 5

1. parallel
2. perpendicular
3. perpendicular
4. parallel
5. parallel
6. perpendicular
7. perpendicular
8. parallel

to a disk-o

Page 6

1.

   A   B
   R   S
2.

   C
   T   U
   D
3.

   M
   G   W   H
   N
4.

   X
   U   V
   S   Y   T

Page 7

1. skew
2. intersecting
3. intersecting
4. intersecting
5. skew
6. intersecting
7. skew

Indiana

Page 8

1.
2.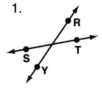
3.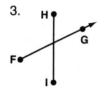
4.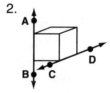
5.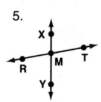
6.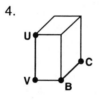
7.
8.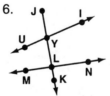

Page 9

E. 155°	I. 25°	E. 180°
N. 205°	T. 310°	L. 50°
H. 130°		

the Nile

Page 10

Answers may vary.

1. ▱EFG, ▱E
2. ∠ABC, ∠CBA
3. ∠R, ∠2
4. ▱STU, ▱S
5. ▱XYZ, ▱W
6. ∠D, ∠1
7. ▱NMO, ▱P
8. ▱STU, ▱V
9. ∠CAB, ∠BAC
10. ▱FGH, ▱I

Page 11

1.
2.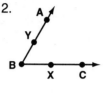
3.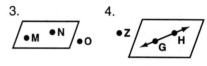
4.
5.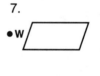
6.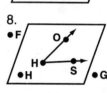
7.
8.

Page 12

Check students' constructions.

Page 13

Check students' constructions.

Page 14

1. 135°, obtuse
2. 30°, acute
3. 90°, right
4. 113°, obtuse
5. 90°, right
6. 125°, obtuse
7. 36°, acute
8. 90°, right
9. 116°, obtuse
10. 150°, obtuse
11. 42°, acute
12. 117°, obtuse

6 years old

Page 15
1. 60°   2. 135°   3. 25°
4. 110°   5. 45°   6. 55°
7. 90°   8. 115°
lawsuits

Page 16
1. ∠1 and ∠2, ∠2 and ∠3, ∠3 and ∠4, ∠4 and ∠8, ∠8 and ∠7, ∠7 and ∠6, ∠6 and ∠5, ∠5 and ∠1
2. ∠1 and ∠8, ∠2 and ∠7, ∠3 and ∠6, ∠4 and ∠5
3. 40°
4. ∠2 and ∠3
5. 80°
6. vertical = ∠6, adjacent = ∠2 and ∠4
7. 50°
8. ∠8, ∠6, and ∠2

Page 17
1. 90°   2. 40°   3. 50°
4. 180°   5. 130°   6. 0°
7. 10°   8. 220°   9. 360°
blue whale

Page 18
Check students' constructions.

Page 21
1. 360°   2. 540°   3. 1080°
4. 720°   5. 180°   6. 1800°
7. 2520°   8. 3240°
He made his bed!

Page 22
1. pentagon: 5, 5, 5
2. dodecagon: 12, 12, 54
3. hexagon: 6, 6, 9
4. triangle: 3, 3, 0
5. quadrilateral: 4, 4, 2
6. decagon: 10, 10, 35

Page 23
1. quadrilateral   2. hexagon
3. triangle   4. dodecagon
5. pentagon   6. decagon
7. octagon   8. heptagon
9. nonagon
a doughnut

Page 24
1. not a polygon
2. not regular
3. regular
4. not regular
5. regular
6. not a polygon
7. not regular
8. regular
Simpsons

Page 25
1. concave
2. concave
3. convex
4. not a polygon
5. convex
6. not a polygon
7. not a polygon
8. concave
9. not a polygon
on grillas

Page 26
1. octagon   2. pentagon
Pictures could vary.

3. decagon   4. hexagon
Pictures could vary.

5. heptagon   6. quadrilateral
Pictures could vary.

Page 27
1. 2   2. 4   3. 0   4. 1
5. 2   6. 6   7. 2   8. 3
9. 1   10. 0
21 years old

Page 28
1. 180°   2. 180°   3. no
4. 90°   5. no   6. 45°
7. 180°   8. 45°
two

Page 29
1. $\overline{FG}$ 2. $\overline{IH}$
3. ∠J   4. ∠H
5. 12 cm   6. 4 cm
7. 135°   8. 150°
9. $\overline{FG} + \overline{HI} + \overline{JF}$
10. m∠A + m∠C + m∠D

Page 30
M. a = 10, b = 22, c = 90°
U. a = 15, b = 7, c = 6
J. a = 10, b = 13, c = 75°
R. a = 19, b = 55°, c = 130°
A. a = 108°, b = 12, c = 108°
Y. a = 12, b = 16, c = 25
A. a = 4, b = 20, c = 10
Jam-uary

Page 31
1. 90°   2. 40°   3. 108°
4. 120°   5. 50°   6. 135°
7. 75°   8. 15°
Mount Everest

Page 34
1. 55°   2. 60°   3. 45°
4. 90°   5. 125°   6. 155°
7. 50°   8. 20°   9. 10°
They get sentenced!

Page 35
1. scalene   2. equilateral
3. isosceles   4. scalene
5. isosceles   6. equilateral
7. scalene
Beatles

Page 36
1. acute   2. right
3. obtuse   4. acute
5. acute   6. obtuse
7. right   8. acute
2 pounds

Page 37
1. ∠2 and ∠3   2. ∠1 and ∠2
3. ∠1 and ∠3   4. 65°
5. 145°   6. 100°
7. 115°   8. 35°
9. 145°   10. 115°

Page 38
1. SAS   2. SSS   3. ASA
4. SSS   5. SAS   6. ASA
7. SSS

the sink

Page 39
Check students' drawings.

Page 40
Check students' constructions.

Page 41
1. A, C, E, G, J
2. A, B, J
3. A, C, E, F, G, H, I
4. A, C, D, E, F, G, H, K
5. A, C, D, E, G, L

Page 42
1. false   2. true   3. true
4. true   5. false   6. false
7. false   8. true   9. false
10. false

six

Page 45
1. 60 m   2. 99 m   3. 88 m
4. 48 m   5. 54 m   6. 52 m
7. 149 m   8. 128 m

Ciderman

Page 46
T. 48 in   I. 36 in   N. 42 in
E. 62 in   F. 46 in   E. 45 in
F. 33 in

fifteen

Page 47
1. 43.96 m   2. 25.12 m
3. 31.4 m   4. 18.84 m
5. 37.68 m   6. 50.24 m
7. 47.1 m   8. 62.8 m
9. 34.54 m

loose leaf

Page 48
1. 50.2 m   2. 22.0 m
3. 28.3 m   4. 53.4 m
5. 65.9 m   6. 59.7 m
7. 100.5 m   8. 27.0 m
9. 40.8 m

New Jersey

Page 49
1. 16 in   2. 35 ft
3. 10 in   4. 5.5 cm
5. 25 ft   6. 7 in
7. 7,926 mi   8. 2 in

Page 52
1. 875 ft^2
2. 162.5 in^2
3. 2850 cm^2
4. 1500 ft^2
5. 26.25 mm^2
6. 67.5 ft^2
7. 49.49 mm^2
8. 720 in^2

Page 53
N. 169 m^2 I. 1750 m^2
L. 132 m^2 R. 1280 m^2
A. 195 m^2 F. 256 m^2
O. 121 m^2 A. 627 m^2
C. 30 m^2 I. 12 m^2

California

Page 54
1. 65   2. 276   3. 126
4. 164.5   5. 900   6. 146
7. 512   8. 636

Dino-sore

Page 55
1. 65   2. 777   3. 144
4. 416   5. 654   6. 32
7. 46   8. 16

2,150 miles

Page 56
1. 195 m^2 2. 250 mm^2
3. 112 ft^2 4. 1125 cm^2
5. 792 in^2 6. 2000 yd^2

Page 57
1. 153.86 in^2
2. 28.26 in^2
3. 254.34 in^2
4. 452.16 in^2
5. 1591.98 in^2
6. 307.72 in^2

Page 58
1. 180 m^2 2. 178.5 m^2
3. 506 m^2 4. 1050 m^2
5. 108 m^2 6. 379.94 m^2
7. 169 m^2 8. 907.46 m^2

rep-tiles

Page 59
1. 47.44 m^2 2. 117 m^2
3. 225 m^2 4. 154 m^2
5. 289.94 m^2 6. 115.5 m^2
7. 763 m^2 8. 536.74 m^2
9. 417.62 m^2

Bill Gates

Page 62

Shape	Sketch of Shape	Number of Faces	Number of Vertices	Number of Edges
triangular pyramid	See	4	4	6
triangular prism	students'	5	6	9
rectangular pyramid	sketches.	5	5	8
rectangular prism		6	8	12
pentagonal pyramid		6	6	10
pentagonal prism		7	10	15
hexagonal pyramid		7	7	12
hexagonal prism		8	12	18

Page 63
1. rectangular prism
2. triangular pyramid
3. cylinder
4. triangular prism
5. sphere
6. cone
7. rectangular pyramid
8. hexagonal prism

3 touchdowns

Page 64

E. 384 m²   T. 2088 m²
A. 270 m²   I. 612 m²
M. 448 m²   D. 4672 m²
C. 696 m²   E. 2340 m²

miced tea

Page 65

1. 125 cm²   2. 245 in²
3. 504 m²   4. 896 ft²
5. 1584 yd²   6. 2580 mm²
7. 301.75 in²   8. 1968.75 m²

Page 66

1. 351.68 m²
2. 979.68 m²
3. 1526.04 m²
4. 1657.92 m²
5. 226.08 m²
6. 2920.2 m²
7. 4747.68 m²
8. 4163.64 m²

Kentucky

Page 67

1. 846 cm²   2. 2970.44 in²
3. 216 ft²   4. 864 m²
5. 12,246 in²   6. 583 mm²
7. 1332 yd²   8. 3336.25 in²

Page 68

1. 395.64 m²
2. 502.4 m²
3. 1909.12 m²
4. 213.52 m²
5. 406.63 m²
6. 2712.96 m²
7. 3456 m²
8. 2590.5 m²

the mealman

Page 69

1. 452.16 in²
2. 1017.36 in²
3. 2826 in²
4. 5024 in²
5. 7850 in²
6. 254.34 in²
7. 1519.76 in²
8. 9847.04 in²

Star Wars

Page 70

1. 593.46 in²
2. 706.5 in²
3. 1017.36 in²
4. 1519.76 in²
5. 1648.5 in²
6. 2826 in²
7. 753.6 in²
8. 7850 in²

Sphereheads: 12,902.24 in² to 4012.92 in²

Page 71

1. 576 m³   2. 56 m³
3. 3600 m³   4. 1056 m³
5. 15,000 m³   6. 273 m³
7. 4096 m³   8. 594 m³
9. 28,800 m³

Comma down!

Page 72

O. 9043.2 m³
I. 24,316.16 m³
E. 942 m³
G. 14,469.12 m³
R. 43,960 m³
A. 4615.8 m³
G. 38,659.68 m³

Georgia

Page 73

1. 7234.56 in³
2. 615.44 in³
3. 14,130 in³
4. 11,539.5 in³
5. 24,416.64 in³
6. 12,560 in³
7. 57,876.48 in³
8. 38,151 in³
9. 38,772.72 in³
10. 76,930 in³

ten

Page 74

1. 1890 in³
2. 360 m³
3. 10,048 mm³
4. 6079.04 ft³
5. 24,416.64 cm³
6. 24,000 m³
7. 448 ft³
8. 38,772.72 mm³
9. 19,292.16 in³
10. 68,687.5 cm³

FS122011 Introduction to Geometry Made Simple ▪ © Frank Schaffer Publications, Inc.